Anonymous

Pocket companion for locomotive engineers and firemen : containing general rules and suggestions for the management of an engine under all circumstances

Anonymous

Pocket companion for locomotive engineers and firemen : containing general rules and suggestions for the management of an engine under all circumstances

ISBN/EAN: 9783337156626

Printed in Europe, USA, Canada, Australia, Japan

Cover: Foto ©berggeist007 / pixelio.de

More available books at **www.hansebooks.com**

POCKET COMPANION

FOR

LOCOMOTIVE ENGINEERS AND FIREMEN:

CONTAINING

GENERAL RULES AND SUGGESTIONS

FOR THE

MANAGEMENT OF AN ENGINE UNDER ALL CIRCUMSTANCES.

BY CHARLES A. HOXSIE,
PRACTICAL ENGINEER.

FIFTH EDITION.

ALBANY:
WEED, PARSONS AND COMPANY, PRINTERS.
1876.

Entered according to Act of Congress, in the year 1875, by

CHARLES A. HOXSIE,

In the Office of the Librarian of Congress, at Washington.

CONTENTS.

	PAGE.
Introduction	5
The Locomotive	9
The Fireman	11
The Engineer	17
Advice to young Engineers	23
Tramming and Center-marking	30
Adjusting Side and Main Rods	32
Pumps and Pump Valves	38
The Cylinder and Cylinder Packing	44
Valves and Valve Motion	49
Extreme or Dead Center	56
The Eccentric	60
Setting Eccentrics	63
Trouble on the Road—Four Points of Valve Motion	66
Pumping on the Road	69
Counter-balancing	72
Expansion and Expansion Braces	73
Accidents and Temporary Repairs	75

CONTENTS.

Miscellaneous Suggestions - - - -
General Remarks - - - - -
Special Advice to Engineers and Firemen -
Index - - -

INTRODUCTION.

This little work is designed for Locomotive Engineers and Firemen. It is written with a view of aiding, in a spirit of earnest sympathy, those desirous of fully understanding and mastering their responsible vocation. While there are not many works that explain, plainly and thoroughly, and with freedom from unnecessary technicality, the management of the locomotive, it is true that in no other branch of mechanical industry is a correct knowledge of principles and details more important. Indeed, the possession of such a knowledge is an indispensable requisite to the man who is daily intrusted with the safety of human life and property. Believing, from a practical experience of many years, that there is positive need of a manual that will impart to Engineers and Firemen precise practical knowledge on the subject of their duties, and the general care of an engine, the author has been induced to prepare the treatise which follows. He does not claim for his work a high degree of literary excellence, but he has the vanity to believe that in putting upon paper the actual results of and deductions from his own experience, he has placed it within the power of any competent and intelligent mechanic to become entirely familiar with his duties, and, in fact, to make himself a first-class Engineer.

The book is designed also as a pocket companion for those as well who have mastered their profession, much embarrassment being often avoided by the possession of a convenient work of reference when exact information is desired. It is anticipated, therefore, that the work will prove as valuable to the experienced Engineer as to the tyro in the profession. That it will have a tendency to elevate the Engineer in his calling and add to his usefulness to his employer and the public is the earnest hope of the writer.

In placing the result of his labors before his brother Engineers, the author begs leave to state that he has endeavored throughout to conform himself strictly to the matter in hand, viz., the succinct explanation of the use and operation of all the important parts of the locomotive engine, its care, and the proper performance of the duties devolving upon the Engineer. The explanations and instructions will be given in the plainest possible language, devoid of scientific terms, and condensed as much as is consistent with entire clearness of statement. The reason of things as well as the operation and the effect, will be fully explained, and no idea will be advanced that cannot be entirely substantiated. In short, it is designed that it shall constitute a comprehensive and common sense manual of locomotive engineering, and as such it is confidently hoped that it will meet, in an especial manner, the necessities of those who have not had an opportunity to acquire that knowledge which their profession calls for. The book is not designed for the instruction of locomotive builders or machinists, but solely for the improvement of Engineers and Firemen, and therefore much that is merely technical in regard to the construction of the engine is omitted. Should it prove a useful and valued companion to those ambitious to excel in the line of honorable effort, the author's object will be fully attained.

POCKET COMPANION

FOR

ENGINEERS AND FIREMEN.

THE LOCOMOTIVE.

A locomotive engine may be defined as simply two engines connected together by working from one driving shaft, operating also in conjunction by being situated at right angles with each other, in order that one may assist the other in passing the center points of the crank. The fact that its office is the furnishing motive power for the transportation of freight and passengers from one part of the country to another, indicates that the greatest care is demanded in its operation. The principles of its construction and the relation of the parts one to another should be fully understood by those who would essay its management. In proceeding to explain those principles and relations, we find that there must be a starting point in entering upon the work of progression from the apprentice o the thorough Engineer. This starting point will be the Fireman's post on the foot-board, as no person can be a competent Engineer until he has learned, in that capacity, to properly take care of an engine. Consequently it is with the duties of the Fireman that we have first to do, and it is in their faithful and intelligent performance that those habits of neatness and system are learned, which are essential to the efficiency of the Engineer.

THE FIREMAN.

It devolves upon the Fireman, previous to starting upon a trip, to place the engine in complete readiness for service, clean every thing thoroughly, and kindle the fire properly and expeditiously. His first duty, however, before starting the fire, is to see that there is water in the boiler, and it is of paramount importance that this duty should not be neglected. The match should not be applied until there is at least one gauge of water in the boiler. Two gauges are better than one, but one will be safe.

The neglect of this duty will inevitably result in the destruction of the crown sheet and flues, occasioning serious pecuniary damage as well as injury to the reputation of the heedless Fireman. Hence the first thing to be done is to try the gauges, and they being right, to start a very slow fire until the flues and smoke stack get warm and expansion begins by degrees. The warmth will increase the draft, and the smoke will thus pass off with ease.

If an excess of oily waste be used, or the fire-box be filled with wood before the fire is properly started, the cold flues will not carry off all the smoke, which will naturally escape around the door with unpleasant results, besides choking the fire.

The tank should be filled with water as soon as a fire is started, provided the facilities are at hand. If otherwise, it should be the invariable rule to procure water and fuel the first thing after leaving the house. This duty performed, the Fireman should see that sufficient oil and waste is on the engine for the trip. The interior of the cab should be thoroughly cleaned, and if it be a night trip the lamps should be made ready. The cab is the home of the Engineer and Fireman during a large portion of their time, and thorough cleanliness should be the rule with every portion of it.

In oiling the engine care should be taken not to waste the oil. A little should be placed on the tops of the wedges to prevent them from sticking, and a squirt-can should always be used to oil the small work. In performing this duty there should be a sharp watch for loose or absent nuts or stopped up oil-holes, all of which should be promptly reported to the Engineer. Diligence in these matters, as well as in being about the engine in ample time previous to a trip, will indicate the interest felt by the Fireman in his work. Habits of neatness, order and punctuality should be systematically adhered to.

The engine being in readiness for the start, the condensed water should be allowed to escape from the cylinders by opening the cylinder cock, and when the engine is moved, it should be very slowly, until all the condensed water is freed. If this is not attended to, the packing will be forced down, and made to blow, and the water will come out of the stack and smear the engine. In running over the road the Fireman should at all times be watchful, not only in keeping up the fire, but in attending to the Engineer's instructions. Firing should not be done while going through or into stations, or crossing bridges, or turning short curves, as a good look-out is especially essential at these points, and this cannot be maintained if the attention is occupied with firing. In the night, particularly, the light from the furnace will greatly interfere with the sight of the Engineer, while if any thing should happen the Fireman would be in no condition to care for himself.

On arrival at the terminus, the Engineer usually leaves the engine in charge of the Fireman, who should be careful in putting it away that every thing is right after it is placed on the wheel. In running into the house the engine should be so placed that the smoke-stack will stand exactly under the smoke openings, then he should shut the throttle off entirely, set up the thumb-screw, if any, put on the tender brake with sufficient force to hold the wheels in case of an elevation in the track, open the cylinder cocks and leave them open, place the reverse lever in the center notch of the quadrant, see that the scales are properly adjusted, if necessary, to prevent an excess of steam when the engine is fired up again, and shut the dampers and furnace doors, and also the slide if there is one. After satisfying himself that the fire is all right, if any is left, the Fireman should then open all the gauge-cocks, one at a time, for the purpose of freeing them of sediment. If no water is seen when the steam is all off, the throttle should be opened to destroy whatever vacuum there may be, when the water will show itself if there be any. All the oily waste should be picked up and put away, and the tools put in their proper places. There should be a place for every thing, and every thing in its place, so that when the Fireman leaves the engine he can feel assured that all will be right when he returns. Railroad companies sometimes employ men for the purpose of caring for engines brought in from the road. They are usually termed hostlers, and are generally selected from Firemen of known reliability. The attainment of the position is regarded as an advanced step in the line of promotion. Those Firemen, therefore, who hope for advancement, should not only make themselves thoroughly familiar with their duties, but aim to establish a reputation for integrity and attention, losing no opportunity to fit themselves for a higher rank in their avocation. It is not necessary that they should restrict themselves to the precise duties required by the company, though these should at all times be faithfully performed, but much will be learned by assisting the Engineer, and cultivating an enquiring turn of mind. The information thus gained will amply compensate for the time and labor involved in obtaining it, and the certain result of diligence and application, in this respect, will be that they will ultimately become capable of assuming an Engineer's responsibilities. They should bear in mind that at no time is the preliminary knowledge of the Engineer so well obtained as while he fills the subordinate position of Fireman, and it is certain that promotion will find the man who does not possess that knowledge in a position of very great embarrassment. Asking questions will not be very pleasant, and many young Engineers would, perhaps, rather remain in ignorance than expose their shortcomings.

THE ENGINEER.

Supposing the reader to be thoroughly familiar with the Fireman's duties, and promoted to the more responsible post of Engineer, we will now have something to say as to his duties in the new capacity.

The first, and most indispensable requisite, to the man who assumes the Engineer's responsibilities, is a good and valuable time-piece. Not until he has that article in his pocket is he ready to go upon his engine. When he has taken his place he should satisfy himself through the Fireman that the tank is full of water, that he has sufficient fuel to make the trip, and that there is a full complement of tools on board. These should include a hammer, screw wrench, other wrenches for packing, chisels, two large jacks, jack levers, two chains, pinch bar, an axe and a saw. With these in their proper places, and with steam up, he is ready to proceed to the depot or freight yard for his train. On arriving there the standard time should be obtained, and a time card procured, the instructions upon which should be well studied. He is then ready to make up his train, if he has that work to do. In performing this task great carefulness should be exercised, that no accidents or casualties occur. The start should be made with care, avoiding as much as possible sudden jerks to the train, keeping meantime a good lookout to the rear in order to be certain that the entire train follows the engine. The run over the road should be made with steadiness and circumspection, in all cases adopting a reduced rate of speed in approaching and passing stations and meeting places, watching the train closely in turning curves, and in general observing the requirements laid down in the company's code of regulations, as far as it is possible to do. In running under telegraphic instructions, Engineers should never leave a station until the order is thoroughly understood.

In this connection there are several points which should be treated of more fully, and it is necessary to go back once more to the starting point. It is of the first importance, especially in coal burning locomotives, that enough water should be in the boiler to allow of leaving the pump off until the fire gets well started, and the steam gets up to the standard point. If this be not attended to, and the water is low, perhaps down to one gauge, and the fire still burns slowly, the steam will run down quickly, even though it be up to the standard when the start is made, considerable time elapses before a full fire is obtained, and before the steam is again up, the pump must be brought into requisition to keep the water up, which again reduces the steam, and if the Engineer is not "stalled" he will certainly lose a great deal of valuable time. A little due preparation at the right time will therefore obviate considerable trouble. With three gauges of water, or thereabouts, the Engineer is enabled to shut off the pumps and give the fire a chance to burn, while steam is made all right with plenty of water in the boiler. The pumps may then be set to the regular feed, and there is no trouble. Every thing will go well, when matters are managed as they should be.

After the start is made the engine should be worked expansively by pulling back the reverse-lever, and working in that way to the full extent compatible with making the running time. By doing this less fuel and water will be used and will make better steam with less labor to the engine in running to time.

The Engineer should at all times keep his mind upon his business, particularly as regards his meeting point, in case he is running on a single track road, and should remember Artemas Ward's truism that "one train trying to pass another on a single track always results in failure." He should allow himself ample time to reach the meeting point, taking into consideration the condition of the engine, the character of the train, the grade of the road, the condition of the rail, etc., and in case of doubt, it is always well to take the safe side. No long runs for water should be made, as there is never any reasonable excuse for being caught without water, when suitable provision has been made for a supply at convenient distances. Should it become necessary from any cause to let the fire go out in freezing weather, the pumps and feed pipes must be carefully attended to by taking off the hose and breaking joints to the pumps to let out the water, taking care also to raise the top air chamber, and place a piece of wood in the joint to keep it open to prevent the freezing of any leakage from the check-valve.

Whenever the Engineer is in trouble he should in all cases avoid blocking the road if possible.

On the completion of the trip the Engineer should calculate to have plenty of water in the boiler by the time he is ready to put the engine away, avoiding excessive pumping after leaving the train. Before leaving the engine he should look it over and see that every thing is all right for the next trip. The brake should be let off, and the tender wheels sounded with the hammer, the springs and truck of the engine and tender should be examined for loose or absent nuts and bolts, and it should plainly appear from actual observation that every thing about the locomotive is in its proper place and condition. A close watch ought to be kept upon the flanges of the wheels that they do not get worn too sharp and become dangerous in passing frogs. When they become worn they should be promptly reported and taken out without delay. The flanges on the driving-wheels should also be watched. When the engine gets out of tram the tires will shoulder up and become sharp. To avoid that the Engineer must tram the driving-wheels. If no report is made of defects it will be supposed that the engine is all right, and should the master mechanic want an engine in the night, and taking one under that supposition find it in need of repairs, it would not be creditable to the Engineer having charge of it. It therefore behooves Engineers to be alert and watchful in this respect, if they would deserve and receive the credit due to careful performance of duty.

The Engineer's position is thus one of continued and onerous responsibility, and it is not possible for him to become too thoroughly familiarized with details. A reputation for efficiency is not obtained without persevering effort.

ADVICE TO YOUNG ENGINEERS.

A word of advice to the young Engineer just entering upon his duties cannot be amiss, as oftentimes a good reputation is won or lost at the very start. The Fireman who is steady, minds his own business, aims to promote his own as well as his employer's interest by faithfulness to duty, and acts the part of a man in all his dealings with his fellow men, is worthy of promotion. If he fails in any of these points he is not worthy. When he is promoted, he must keep his feelings in reference to his elevation within the bounds of good judgment, and never allow himself to become overbearing or too independent. His mind should be concentrated upon his profession, and he should understand and appreciate the fact that he has a responsible part to perform, and determine to perform it like a man. He may be obliged to work many hours when other Engineers are apparently at leisure, but he should do it cheerfully, performing every duty conscientiously, and learn to care for his engine with painstaking solicitude. It is not well to attempt to run faster than any body else. It is sufficient to make reasonable time. With a watchful eye to business, the Engineer should be kind to every body, and use respectful language at all times. He should never allow himself to get into a passion. or by cursing and abuse provoke the ill-will of those brought in contact with him, but he should so conduct himself as to secure the respect and regard of all, and thus render his daily duties pleasant, establishing meanwhile a lasting and honorable reputation.

When a young Engineer is placed on a strange engine, or one that is old, loose and about used up, he should never key up the rods until he has run one or two trips, and ascertained about where the lost motion is. The wedges should be set up all around, but not tight enough to stick. Then place the engine on the center forward and back, and key up the rods, leaving them loose enough to prevent them from running hot. Look to the lubricators, or fenders on the rods, and see that they have wicks in them, and are all right, and feed freely. In reference to setting the wedges and keying the rods, it is better to adjust them twice over than have the wedges stick or the pins cut. There is a great difference in engines in this respect, and, if the work is done by degrees, it will cost less trouble in the end. When the rods are keyed up, it is well for the Engineer to let some one move the engine ahead while he is trying the rods. If they shake, it is an indication that they will not run hot. If they are firm, let up the key until they can be shaken. Sometimes the rod will be loose at one point and tight at another. Put the engine on the tight point, and let up the key until it is loosened. It is better to have all the lost motion in the back end of the side rods, and not have both ends loose and rattling. It will be impossible to get some engines exactly right, and about all that can be done, as to the working parts, will be to see that the wedges and rods work perfectly. The Engineer should also closely inspect the main box feeders, see that the oil holes are clear, and that the cellar is properly packed. The engine and tender trucks should also be examined.

There will always be something to do when there is a leisure moment. When the rods bother, get them as near right as possible, and try them when the engine is working hard and slow. If they are loose at all points, they cannot be bettered except by keeping them keyed up snug.

By trying different plans, and finding where the lost motion is, new ideas will be learned that will ultimately be of value. If it be found, after the engine is all keyed up, that there is a pound somewhere, place it on the quarter stroke, and block the driving wheels. Then, by using a little steam, and working the reverse lever backward and forward, the Engineer can watch the side rods or main rods, and the main box, and ascertain where the trouble is. He must not be discouraged if the engine works badly, but ask the advice of more experienced Engineers, and keep on trying without complaining of the hard luck. The lesson will ultimately be a good one, and will not only strengthen the faith of the tyro in his own capacity, and give him substantial encouragement, but his superiors will note his struggles, and think the better of him for his efforts to master his difficulties. It may be up-hill work at first, but it must be remembered that no honors are won in any profession without hard work.

The engine should always be moved around yards and stations with the utmost care and watchfulness. A good lookout should be kept at branches, and when leaving stations. It will be time enough to get satisfaction out of the engine when there is plain sailing ahead. The open road is a better place to show ignorance than the yard or station, and nothing will be more hurtful to the reputation of the beginner on the foot-board than leaving stations at a reckless rate of speed.

Young beginners are usually placed on freight trains at first, and it may be that the first runs will be made in the night, but whether it be in the day or night-time, a close watch must be kept on the train that no breaks occur. The Engineer or his Fireman should look back at every curve, and when the start is made he should be certain that the entire train follows. It does not look well to see an Engineer running fifteen or twenty miles with but half of his train. It is not safe to depend upon the bell-cord.

On most roads it is the practice to employ men for the purpose of packing trucks. This is all well enough, but it is nevertheless the Engineer's duty to inspect the engine and tender journals, main boxes, etc. Serious trouble has often been occasioned by running trains with blazing journals, running them sometimes until they break off. It is a safe practice never to run by a station with a hot journal without remedying it.

It is generally the case when a Fireman is promoted, that his first essay is upon a poor or worn out engine, sometimes the worst the company has, and the young Engineer is given to understand that he is expected to do good work with it. Often the engine dispatcher will inform him that it will do good work when there is really considerable opportunity for doubt upon the subject. This unquestionably places the young Engineer in an embarrassing position, and it is very sure to show of what sort of stuff he is made. Indeed, it is usually the case that he is assigned to the poor engine with that very object, and if he succeeds in doing good work with it, it constitutes positive evidence that he is able to run a good engine. The man who is placed in such a position should not therefore be discouraged. He must make up his mind to do his best, and the victory will be worth the winning.

TRAMMING AND CENTER-MARKING.

What is technically termed "tramming" consists in so adjusting the driving wheels that they shall stand square with the frame of the engine. It is generally done when repairs are being made in the shop, but it is important that every Engineer should be familiar with the most expeditious mode of doing it. When the driving wheels are taken from under the engine, the distances are equalized both ways from the face of the blind wedge of the forward main box to the male casting which is fastened to the smoke box, between the cylinders, and in the center of the boiler. After the blind wedges to the forward driving boxes are trammed, all the driving boxes being ready, they are placed in the pedestal or jaws of the main boxes, and the center of the boxes being obtained, the back wedges may be adjusted by tramming from the center of the forward driving box to the center of the back driving box, thus equalizing the distance of both main centers on both sides of the engine.

TRAMMING AND CENTER-MARKING. 31

Much trouble is often avoided by having an engine properly center-marked. This is done as follows: After the blind wedges to the forward boxes have been trammed and put in place, then tram from the face of the blind wedge of the forward driving box to the outside of the pedestal or jaw to which the blind wedge is fastened. Tram as far back from the face of the wedge as possible, and make a prick punch mark on the pedestal, using the same tram for both sides of the engine, taking care that the distances from the face of the wedges to the marks on the jaws are alike on both sides. Tram from the middle of the jaws. When engines get out of tram, as they do frequently, the center marks may be used to adjust the machine without raising it off the driving wheels. The forward driving wheels should be trammed first from the center marks and then made to serve as a guide for the rear drivers. Particular attention to this matter is essential to secure the easy working of the locomotive and the avoidance of hot crank-pins and similar drawbacks.

ADJUSTING SIDE AND MAIN RODS.

The proper adjustment of the side rods is a somewhat difficult task, and requires the exercise of considerable patience on the part of the Engineer who is careful to get them exactly right. It should always be done while the boiler is under a pressure of steam. Sometimes it will be necessary to put them up and take them down several times before they are satisfactorily adjusted. Before placing them in position, the main centers should be trammed from the center marks. After setting up the wedges snug, a few trips should be run in order to ascertain whether or not they are so tight as to require letting down, which should be avoided if possible after the rods are adjusted. The wedges being right, the side rods should be taken down and the pins examined to see whether they do not require re-turning. The side rod brasses should then be reduced to the pins, the brasses when keyed up being left brass and brass. The straps should then be adjusted with the brasses in their places, the latter keyed up tight, and then the exact center of each brass on both ends of the side rods should be obtained. The side rods being thus ready for tramming, the main centers should be trammed alike on both sides, equi-distant from center to center, and the same tram may be used on the pin centers, and should they not be exactly right, as is often the case, the rear driving wheels should be slipped until they come right. The main centers and the pin centers being alike, the side rods may then be trammed with the same tram, applying it from center to center of the brasses, putting in or taking out "liners" to bring them exact. The key should then be driven down hard and marked next the strap with a scribe or knife. When this is accomplished on both sides and the distances between main centers, pin centers and brass centers are exactly the same, the side rods may be put up, driving the key to the mark made while tramming, and the work is then properly done.

Frequently a side rod will get too short or too long owing to unequal wearing or keying of the brasses. It is then essential that the rod should be adjusted to the proper length, and before it is taken down the end which remains in position should be keyed up. In adjusting the right hand rod, the cross-head should be placed on the extreme forward center, when a screw-jack may be placed for convenience sake under the strap end of the rod and the strap slipped back so that the liners back of the brasses may be reached, when they may be lessened or increased as required. There should be as many liners back of the inside brass as it will take without wedging the pins apart, while as many should be placed behind the strap brass as will permit the bolts and key to occupy their proper position through the strap and rod. The rod should then be keyed up as tight as possible, shaking it with one hand. In the case of an old engine, or one that has been out of the shop from twenty to thirty-six months, it is important that the cross-head should be placed on the extreme forward center on the side of the engine from whence the rod is removed. In such an engine, the pins are likely to be worn out of round where

ADJUSTING SIDE AND MAIN RODS.

the friction is greatest, and that point is from the forward side of the main pin to the following quarter. When a rod is keyed up it should always be keyed on the largest part of the pin, so that when the engine is moving the brasses will not bind upon the pin, and the rod will then run cool and give no trouble.

A word or two is also proper in reference to adjusting the main rods. In all cylinders there is, or should be, a clearance of from one-quarter to three-eighths of an inch, to prevent the piston from striking the cylinder-heads as the rod varies in length by reason of wear or from unequal lining and keying of the brasses. The clearance should be equally divided in both ends of the cylinder in the following manner: First, key up the main rod, that all lost motion may be taken up, then ascertain the extreme travel of the piston at both ends of the stroke by placing the cross-head on the extreme center forward and back, making a mark across the guides and cross-head. These marks will be the traveling points of the piston. The striking points of the piston should then be obtained by disconnecting the main rod, prying the cross-head back until the piston strikes the cylinder-head and making a mark from the mark first made on the cross-head across the guides, which will be the striking point at the back end of the cylinder. Then pry the cross-head forward until the piston strikes the forward cylinder-head, and a mark as before will constitute the striking point at the forward end. The difference between the traveling point and the striking point, will be the clearance of the piston in the cylinder, and it should be divided equally in both ends by taking out or inserting liners. The use of a jack under the rods when putting on the strap renders its adjustment to the proper length much easier. A rod may be shortened by taking out liners between the rod and the brasses and putting them in the strap behind the brasses, and lengthened by reducing in the strap and putting them forward of the brasses next to the rod. Putting them between the brasses and the rod has the effect only to raise the key. And when the brasses are filed, as is often the case, it is only necessary to insert a thin liner to keep the key raised to its proper place. In reducing brasses the usual rule is to put the same or sufficient thickness in liners back of the brasses as is taken from the brasses, to keep the rod the same length, and also keep the key raised to the proper height. Special care should be taken, as the Engineer will generally find it necessary to use his judgment as to the thickness of the liners. The edges of the brasses should always be rounded off well, no matter how accurately they may be fitted to the pin, otherwise they are liable to run hot. For example, where brasses on two different rods are reduced, one set being rounded and the other left square, it will be found that the former will run the longest and with much less liability to heat. Rounding the brasses is especially advantageous on engine and tender trucks. The attention of Engineers is called to this more particularly, because of the fact that most men, in reducing brasses, do little more than take off the sharp edge. They should be rounded at least one-eighth of an inch on the edge, and three-eighths back from the edge.

PUMPS AND PUMP VALVES.

All Engineers will agree that every engine should be provided with reliable pumps, capable of being worked either at a high or low rate of speed. It often happens that a freight Engineer is called upon to draw a passenger train over the road with an engine that has been accustomed to low speed. Inevitably he will lose time, and, when an explanation is required, complaint will be made that the engine did not make steam fast enough, and the pump did not work. In almost every case, the pump is the sole cause of the lost time and the lack of sufficient steam, for the engine cannot make steam steadily unless the pump works freely and uniformly. Sometimes, also, the engine will fail to make steam for a short distance. In such a case, if the pump be reliable, it may be shut off, so that the steam can be kept up to the standard, enabling the Engineer to make his running time with little trouble. A good pump, in short, renders an Engineer's duty pleasant under nearly all

PUMPS AND PUMP VALVES.   39

circumstances; while a poor one is continually getting him into trouble, damaging his engine and his reputation.

The locomotive pump has three valves, two to force water into the boiler, and one called the check valve, to take the pressure off the branch or injector pipe, and render the pump less liable to get out of order. The capacity of the pump is governed to a great extent by the lift of the valve, which should be so arranged that the lower or bottom valve will receive no more water than the middle and check valves will receive without causing too great pressure on the pump between the valves. When a pump is overhauled, the lower valve should be set with a lift of one-eighth of an inch, the middle valve with three-sixteenths of an inch, and the check valve five-sixteenths of an inch. These proportions will render the pump reliable for any work required of the engine, except for very slow running on a grade, when the left-hand pump should have a lower valve, with a quarter of an inch lift, and the others respectively a scant three-eighths and half an inch. This will give the engine a left-hand pump for a slow rate of speed, and a good pump to fall back upon, should the right-hand pump become disabled from any unforeseen cause. Pumps set in this way will do good service for thirty or forty thousand miles, a fair year's work for an engine.

The pumps should be packed hard enough to prevent leakage, yet not too hard. Some Engineers pack the pumps as hard as possible with the object of making the packing last six or eight months or perhaps a year. This is wrong. When a pump is thus packed, power is taken from the engine, it is liable to get out of line, the plunger wears out of shape, and creases are cut in it, which render tight packing thereafter an impossibility. Packing will occupy four times as much time, and it will not be any better when done, than if it had been packed with reasonable tightness and packed oftener. Whenever a pump begins to leak it should be repacked. To insure reliability, pumps should be packed three or four times a year.

The Engineer should see to it that the engine tank is kept clean at all times, that there are strainers in his hose, and that he has spare strainers and washers to go with them. When his engine is in the shop, he should satisfy himself that the hole in the boiler, through which the water is received from the pump, or check hole, is properly cleaned out. It often becomes corroded, and sometimes it is closed so that the pumps will not work. He should also satisfy himself, in overhauling the pump, that the valve-seat is above the opening in the cage. Occasionally it gets worn below, in which case the valve shows the proper lift by measure, but does not get the opening indicated by the measure, and hence will not work. In some pumps a strainer is connected to the lower part of the lower valve-seat, in the lower chamber. The bottom of this strainer is perforated about one-third of the length, and is, or should be, air-tight above the perforations. If it is not air-tight, the pump will fail to work either partially or entirely. The bolts holding the pump to the frame should be examined for the detection of loosened nuts. When a pump is first put in an engine it is lined up by the guides, and often liners are used, which should be closely watched. The rings inside the barrel and outside of the plunger should not be permitted to wear so that the packing will work through between the plunger and the rings, as it will be likely to get entangled in the cages and cause trouble.

Among the simple matters requiring watchfulness is the management of the pet-cock. When the foot-cock is opened for working the pump, the pet-cock should always be opened to let the air escape. The working of the pump is then known to a certainty, provided the hole in the pet-cock be large enough to throw a perfect stream which can be seen at night. The plugs in the pet and foot-cocks should work easy. These matters seem trifling, and many Engineers doubtless regard them as too trifling for consideration. Nevertheless, their importance cannot be denied. If an Engineer, accustomed to his own engine, is called upon at night to run one that is strange to him, it will often happen that when he starts and opens the foot-cock to start the pumps, and desires to open the pet-cock to test them, he finds that it will not open until loosened by a blow from a wrench or hammer. Then he is unable to see the light spray which issues, gets angry, and finally orders the Fireman to put on the left hand pump. Down the steam will go before the right hand pump can be got in operation, with very little water in the boiler, and the Engineer is in serious trouble, which would have been avoided had the pet-cock been in proper condition. Every Engineer should look to these things with unceasing vigilance. It is the sleepless attention to the smaller matters that goes to make up the aggregate of duty well performed, and the neglect of them as truly marks the careless or incompetent.

THE CYLINDER AND CYLINDER PACKING.

The old fashioned packing rings are perhaps, all things considered, the most reliable cylinder packing in use, and the exceptions to the cases in which they will answer all purposes are very few. Where the cylinder is round and smooth, the packing rings should be turned in the cylinder, and when the packing is set, once in every four or six weeks, it is essential that the packing rings should be kept clean. When the steam begins to "blow through," as it is termed, or pass between the rings and follower-head, the rings should be taken out and cleaned, and the defect will generally be remedied. The blowing may be detected by a close examination when the follower-head is off. Should the cleaning not be effectual, however, it will be necessary to grind in the rings, to do which, the piston must be taken out of the cylinder, and placed in a clamp or vise. The bottom rings are then to be ground to the back follower-head, the top ring to the bottom ring, and the follower-head

CYLINDER AND CYLINDER PACKING. 45

to the top ring. Cylinder packing should be ground as little as possible, consistent with securing a good bearing upon the follower-heads and rings. In order to try them after they have all been ground in, the rings should be placed on the follower-head, and the latter screwed down hard. Then by tapping the rings with a hammer handle it will be ascertained whether they are loose enough to move quite freely. Should they bind they must be ground still more until they work with entire ease. The process requires much care and patience, but neither should be grudged until they are right. Care should be taken not to set the packing out too tight, and the center of the piston should be left in the center of the cylinder. Every time the packing is set out the rings should be examined for indications of blowing through. Attention should also be directed to the springs, spring bolts, nuts, etc. It should be noted that the packing and valves must be kept from "blowing" as much as possible in order to obtain the full power of the steam, for there is where the power is derived. When the packing and valves are tight the full power contemplated by the size of wheel and cylinder and length of stroke is obtained, but where the contrary is the case, the packing does not hold the steam, and hence there is a loss of power. There is also an injury to the draft, and hence a diminution in the amount of steam made. The importance, therefore, of keeping the valves and packing in perfect condition is sufficiently obvious.

To ascertain where and on which side the packing blows, one general rule may be laid down. The blowing usually begins where the cross-head leaves the extreme end of the stroke forward and back, and after the cross-head has traveled from six to twelve inches, the blowing will diminish and sometimes stop. The pressure of steam on the follower-head at the forward and back center is nearer boiler pressure than at any other point, and if the packing springs are weak the rings will close in and admit the steam between the rings and cylinder or between the rings and the follower-head. When the cross-head has traveled back from six to twelve inches, the pressure is reduced so that the springs will force the rings out to their place, and the blowing will

CYLINDER AND CYLINDER PACKING. 47

stop very nearly, if not entirely. Blowing always commences when a train is started, when the engine is working hard and slow. By watching closely, it can be determined on which side it begins. Sometimes in an old engine there will be a shoulder on one or both ends of the cylinder, and when the rings strike it they will close in and the steam will hold them during a portion of the stroke, or until the pressure is reduced so that the rings will be forced to their place by the springs. It will sometimes be the case that blowing will occur elsewhere than in the packing, when even the sound may seem to indicate that the trouble is in the packing. There is no rule to determine where it is at such times, and it is a difficulty which will puzzle the best engineers. It is a very good way to try the packing when the cylinder head is off, by covering the forward steam port and letting the steam into the cylinder through the back port, and opening the throttle when steam is on. That will decide the matter, as if no steam issues from under the rings, and the packing is tight, the blowing is in the valve. In case of doubt, by opening the throttle a little when the engine is placed so that the blow can be heard, and examining the exhaust pipes to see from which pipe the steam issues, it may be determined on which side the blow occurs. The location of the blowing may also be determined by the sound, as close observation will show a very recognizable difference between the noise of the blowing in the valves and in the packing.

VALVES AND VALVE MOTION.

The philosophy of the valves and their motion should be closely studied by every Engineer. It presents many difficulties, and there are really very few engineers who thoroughly understand it. The explanations which follow will give a fair idea, however, of the principles which govern an important factor in the locomotive engine, and they are made sufficiently minute to dispense with illustrative drawings, it being assumed that the reader is at least familiar with the several parts of the locomotive and able to make practical application of the suggestions.

In the valve-seat are three ports, the two end ports being called steam ports. The steam is admitted into the cylinder through these ports alternately, and after it has forced the piston the length of the cylinder, it passes out of the same port through which it reaches the cylinder, passing under the valve, into the middle port, called the exhaust port, and thence out of the exhaust pipe into the open air. On the under side of the valve is a cavity or recess, which should be of sufficient length to cover the exhaust port and both bridges, so that when the valve moves either way it brings one of the steam ports in direct communication with the exhaust port. Through this cavity the steam passes, after being used in the cylinder, the valve moving sufficiently to cause an opening from the steam port under the valve into the exhaust port, and thence into the exhaust pipe. It will thus be perceived that the engine exhausts under the valve. If the valve be placed upon its seat in a central position, it will be found to lap over about an inch on each side of the steam ports. This is termed the end-lap, and its function is to enable the engine to be worked on the expansion; that is, after the valve has cut off the steam, the engine will work, expansively, while the valve is traveling the length of the lap, resulting in a manifest saving of steam. The cavity inside of the valve, it has been seen, just covers the exhaust port and the bridges separating the exhaust from the steam port, or the forward and back steam ports. But the valve more than covers the exhaust ports, lapping over on the bridge between the steam and exhaust ports. This is called the inside lap, whose function is similar to that of the end lap, producing a saving in steam and consequent economy of fuel and water. An inside lap is a disadvantage on some engines, but on freight engines it adds to the facility of working on the expansion more than the end lap calls for, as the valve must travel the length of the inside lap, in addition to the end lap, before the exhaust can occur. The distance the cross-head travels while the valve is traversing the lap is of course governed by the valve gear, which in different engines varies in throw of eccentric, length of rocker arms, radius of link and travel of valve, there being no rule to determine it exactly. Freight engines which run at a speed of from fifteen to eighteen miles per hour, should generally have an inside lap of one-sixteenth to one-eighth of an inch. More than this is a disadvantage, and will render an engine too slow, but with a lap of this length, there will not only be a large percentage of saving, but the engine will go less frequently to the shop for general repairs, as there is less liability of strain, owing to the comparative slowness of its action. Passenger engines, used mainly for speed, should have very little if any inside lap. In a locomotive which makes forty or fifty miles an hour, using six or eight inches of steam, the valve travels to give an opening on the steam-port of about three-eighths of an inch. Hence, the travel of the valve being very short and quick, if there be as much inside lap as on an ordinary freight engine, the lead on the exhaust would be reduced, and the engine would fail to free itself, or become choked, and could not make the required time with a heavy train. Engineers should be largely governed by the way their engine works. If a passenger engine makes steam freely, fast time being required, there should be no inside lap; but if otherwise, a lap of one thirty-second of an inch on each side would be advisable. No passenger engine should have more than that, with perhaps one inch end lap, and five and one-half throw, varied according to length of port, length of travel of valve, and other circumstances.

What is known as the lead on the valve deserves a minute explanation, it being a matter of importance which is much neglected by engineers. For the purpose of this explanation, we will place the right cross-head on the extreme center forward, exposing the steam-chest with the cover off. The valve is then exposed to view. With the cross-head on the extreme dead-center forward, the reverse lever dropped in the forward notch in the quadrant, and the valve set without lead, the latter should be in the act of opening the forward steam-port. The go-ahead eccentric should then be turned until the valve shows an opening of one-eighth of an inch, or whatever space is desired for the lead. This opening is called the lead, for the reason that when the valve is thus set it will be one-eighth of an inch ahead of the cross-head at all points. The lead may be obtained in three different ways: The valve may be cut off on both ends, or the forward side of the forward steam-port and the back side of the back steam-port may be cut out, or it may be obtained from the eccentric. The last mode is by far the more advisable, and, indeed, the only right way to secure it.

In reference to the term extreme or dead-center, it should be stated that the cross-head is used in most cases to obtain a starting point at the extreme dead-center, it being the easiest of access and most convenient in other respects.

The utility of the valve lead above described consists in the fact that when the engine is working, a full port is obtained more quickly with than without it. Without the lead a full port will be obtained when the cross-head has traveled about seven and a half inches at full stroke, and when working six inches, the travel of the cross-head is about one inch. With a lead the full port will be obtained with a little more than half an inch travel when working six inches, and with seven inches at the full stroke. The distance, of course, will vary according to radius of link and length of hooks. It is apparent that when the cross-head travels far enough to enable the steam to get a leverage on the crank, there will be a greater quantity of steam in the cylinder, and it will not only be nearer boiler pressure, but the expansion will be stronger. The engine will therefore do more work and do it easier, and make better time with an eighth of an inch lead on the valve than without it. No engineer who once notes intelligently the difference between the two conditions will be disposed to deny the advantage of this lead.

In a certain sense the lead is equivalent to the lap heretofore described. If the valve be set from the eccentric, the engine being on the dead center, with an eighth of an inch lead, when the engine is moved, the valve will always be one-eighth of an inch ahead of the cross-head at both ends of the stroke, producing the same result as if the valve should be cut off. The valve set without the lead is in the act of opening the steam port. If lead is given to it, it shows an opening, and thus takes off the lap on receiving steam and puts it on in cutting off steam. Hence lead is lap.

EXTREME OR DEAD CENTER.

To ascertain the extreme center of an engine, independently of the lost motion, the engineer should first find the forward dead center on the right side of the engine by taking up all lost motion in the rod by keying it up, and then barring the engine ahead until the cross-head travels within about two inches of the extreme travel forward. Then he should mark across the guides and cross-head, and before the engine is moved make a prick-punch on the wheel-guard, and place a tram fifteen or twenty inches long, one end resting in the prick-punch and the other near the level of the top of the wheel or tire, making a prick-punch on the tire where it comes. Then bar the engine ahead until the cross-head passes the center and travels back past the mark made on the guides, say half an inch, and move back so that the cross-head will travel forward again until the marks upon it will correspond exactly with the marks on the guide. When this is done the tire should be marked as before with one end of the tram in the prick-punch on the guard, and the center of the distance ascertained with a pair of dividers, then with the tram in the mark on the wheel guard, bar the engine back until the opposite end of the tram strikes the center mark on the tire. This will be the extreme center of the crosshead, independent of the lost motion. The engine should always be barred back, so that the cross-head will travel the same way when the marks are made on the wheel, and it will be necessary to get the dead center at both ends of the stroke, and on both sides of the engine, when setting valves or eccentrics, using the same tram. This mode is used when the engine is in the shop or round-house for repairs. The mode of setting eccentrics on the road will be explained hereafter.

In setting valves, it is necessary to get the port marks first, when the steam chest cover is off, and the valve stem disconnected at the back end. A piece of thin sheet tin, about as wide as the port is long, should be placed in the forward steam port, and the valve pushed up until it bears upon the tin. A short tram, about six inches long, should be provided. Then, from a prick-punch mark, made on the valve stem stuffing box, tram to a similar mark, where it may strike on the valve stem. This will be the forward port mark. The back port mark is obtained in a similar manner, using the same tram. With the port marks all right, the valves may be set at any time with this tram, without taking the steam chest off, and they may generally be made square by running them over from these marks. If not, recourse must be had to the dead center. With the engine on the dead center, the same tram may also be used. Try both ends of the stroke, divide the amount of variation from the port marks, and the valve will be square if nothing be sprung out of place. When on the dead center, tram with the short tram from the prick-punch mark on the stuffing box to the port mark on the valve. What it will fall short of reaching the port mark is the lead at that end. Then try the other port mark, and if the lead is the same, the valve is square. If not, the distance must be equalized, and the lead made alike on both ends of the stroke, by putting in or taking out liners in the hooks, or in the back end of the valve stem rod. In all cases, whenever it is desired to ascertain the amount of valve lead, the engine must be placed on the forward dead center for the forward port, and the back dead center for the back port, keeping the reverse lever in the forward motion, when working that motion, and the contrary, when working the back motion, for both ends of the stroke, forward and back. The travel of the valve is not lessened by increasing or lessening the liners. By shortening the valve stem, the travel is changed from the forward to the back end. Hence, if the valve travels too far forward, say an eighth of an inch, a liner of half that size removed from the back end of the valve stem will equalize it.

THE ECCENTRIC.

As a rule, the throw of the eccentric governs the travel of the valve, unless the top and bottom rocker arms should be of unequal length. These arms are usually made the same length, unless there is not sufficient room for the throw of the eccentric, when the upper arm is made the longest. This, however, will give the valve more travel than is called for from the throw of the eccentric. The rule is that the throw of the eccentric should be one-half the travel of the valve, when both rocker arms are the same length, *i. e.*, when the throw is two and a half inches, it will throw two and a half inches ahead, and the same distance back, making five inches.

The go-ahead eccentric is always connected with the top part of the link, it being sometimes the inside eccentric and sometimes the outside. On all indirect motion engines the throw of the go-ahead eccentric follows the crank pin, and on those governed by direct motion it leads. This will be apparent in this way. If the right hand crosshead be on the forward center the crank pin will also be on the forward center. Therefore if the throw of the go-ahead eccentric follows the crank pin, the throw must be on the upper quarter from the crank pin. By moving the engine slowly it will be seen that the crank pin is brought down, and the go-ahead eccentric moves forward. That carries the lower rocker arm forward and the top one back, and the valve travels back, giving the opening to the forward steam port that is required in order that the piston may be driven back. A practical test will perhaps demonstrate the matter more fully. It is plain, also, that if the throw of the go-ahead eccentric follows the crank pin, the back motion eccentric must lead the crank pin when the engine is moving ahead. By reversing the engine the position of the eccentrics will be reversed. The back motion eccentric becomes the go-ahead eccentric, and the latter takes the place of the former.

Young Engineers frequently fail to get satisfactory answers to the question, what is direct and indirect motion? In the former the throw of the eccentric and the travel of the valve are in the same direction and operate at the same time. In the latter they move in opposition. When the throw of the eccentric is forward, the valve travels backward. Nearly all the locomotives used in this country are on the indirect principle.

SETTING ECCENTRICS.

When the eccentric slips on the road, the Engineer owes it to his own credit to adjust it with as little loss of time as possible. There are several ways of doing it, but the following is believed to be the best and most expeditious: To set the go-ahead eccentric, place the cross-head, on whichever side it may be, on the extreme forward center, putting the reverse lever in the back motion and making a mark on the valve stem close up to the valve stem stuffing box. Then put the reverse lever in the same notch in the forward motion that it occupied in the back motion, turn the go-ahead eccentric until the mark on the valve stem comes to the place where it was first made, fasten the eccentric, and the mishap is remedied. The cross-head is placed on the extreme forward center for the following reason: It will be remembered that the throw of the go-ahead eccentric follows the crank, while the throw of the back motion eccentric leads the crank pin. When the cross-head is on the extreme center the throw of one eccentric is up and the other down, bringing the eccentric hooks the same length, while the valve is in the position to give the lead opening to the forward steam port, which must be opened to admit the steam to press the piston back, whether the engine be moving forward or backward. With the hooks the same length and the throw of the eccentric equalized, when the reverse lever is moved forward and back it has no effect on the valve, and if the marks on the valve stem are made from its position when the reverse lever is in the back motion, those marks will certainly be a correct guide by which to set the go-ahead eccentric.

In connection with this matter, there is another idea worth remembering. The radius rods, or eccentric hooks, as they are more commonly called, are connected to the link, one to the top, and the other to the bottom, the link block being stationary. The hook nearest the block governs the travel of the valve. Hence, placing the cross-head on the extreme center brings one eccentric up and the other down, and the length of the hooks and the throw of the eccentric are equalized. The valve, therefore, is not moved, except the slight distance of the lead and lap, when the reverse lever is placed first in the back motion, and then in the forward motion. The distance is so slight that it is not worthy of consideration on the road. The idea will be better understood by a practical demonstration on an engine.

TROUBLE ON THE ROAD—FOUR POINTS OF VALVE MOTION.

Nothing is perhaps more dreaded by young Engineers than the prospect of trouble with the valve motion while running. It is scarcely necessary, therefore, to ask their earnest attention to the following explanations regarding the proper course to be pursued in such cases: Sometimes, when an engine is running at the ordinary rate of speed, it is discovered by the sound that the engine is not exhausting square, and that something is wrong. The matter should be attended to immediately. The throttle should be closed and the speed reduced, so that the train will move about as fast as a man can walk. Dropping the reverse lever in the quadrant, full stroke, open the throttle sufficiently to secure a strong exhaust, and watch the cross-head or the crank-pin on the right side, noting whether there be a good exhaustion on both ends of the stroke. If there is, the trouble is on the other side. It is all important in deciding this question that the engine should be run very slow. After determining on which side the trouble exists, the engine should be stopped and the matter looked after. A thorough examination should be made successively, of the eccentrics, to make sure that all the bolts are in the strap, the nuts in their places and the hooks properly connected; of the link and link block, observing that the throw of the go-ahead is in position to follow the crank pin, and that the throw of the back motion leads the crank pin; and of the rocker arms and rocker box, making sure that the latter is secure in its frame. If these are all right the trouble is doubtless in the steam chest. In examining these points, which constitute the four leading points of the valve motion, nothing should be overlooked. In making the examination, the crank-pin should be always on the top or bottom quarter, in order to give the valve the full travel when the reverse lever is moved forward and back; whereas, if the crank-pin be on the forward or back center it will leave the eccentrics in such position that the full travel of the valve cannot be obtained. If the travel of the valve be obstructed in any way, the Engineer may detect it by placing his hand upon the valve stem while the Fireman works the reverse lever backward and forward full throw. When an engine has a false valve seat held down by studs, these sometimes get loose and work up so that the valve will hit them. This can easily be discovered by the sudden jerk given to the reverse lever, and it may be determined which side it is on in the way just described. These brief directions will generally direct the Engineer to the weak spot in his engine. It is not possible, however, to make rules which will apply to every circumstance, for locomotives will sometimes give out on the road from causes which will baffle the skill of the most experienced mechanics.

PUMPING ON THE ROAD.

In view of the fact that Engineers are sometimes apt to be over confident in respect to their pumps, and run their engines long distances without trying the gauge-cock, ultimately burning the engine, a few words in reference to the subject will not be out of place. In the first place it should be an inflexible rule not to run an engine farther than would be required to use up one gauge of water without trying the gauge-cock. Pumping while going over the road is a very important matter, much more care being required on heavy grades than over a level road, and the ability of the engine to make steam wherewith to climb those heavy grades is largely dependent upon the pumping. The pump should never be worked after the throttle is shut off, for the reason that the steam will fall when the throttle is again opened, and more ground will be lost than is gained by pumping. If the boiler contains one solid gauge of water when the throttle is shut off, it is better to also shut the pump off, and, by so doing, better time will be made than if two or three gauges of water should be pumped into the boiler; and this should never be done unless there is a probability that the engine will remain some time on a branch or in one place.

Pumping so much cold water into the boiler after shutting off the throttle will cause a too sudden contraction of the flues and fire-box, and probably produce a leak. It is always well to carry as much water in the boiler as will enable the engine to work dry steam. Most engines will steam better with the water high, while there are very few engines, especially those burning coal, that will not sometimes, from unforeseen causes, fail to make steam. It often happens that the fire will get the advantage of the Fireman, and, at such times, if there be plenty of water in the boiler, the Engineer has a chance to favor his engine by shutting off the pump.

When the steam begins to fail, it is better to give the engine the benefit of the pump by shutting it off and let the steam come up. Plenty of steam is thus obtained, and there is also a good supply of water. If any is lost it may be gained back by degrees. It is not advisable to wait until the steam gets low, gaining water by light pumping. The Engineer must, however, call into requisition his best judgment, giving strict attention to the peculiar working of his engine in this as in other matters, and where any change is made he should carefully note the effects of the change, and by that means school himself into such a system of managing his engine as will redound to his credit and establish his reputation.

Before we leave this subject it is perhaps necessary to explain why an engine is injured by pumping with the throttle shut off. It is to be noted that when using steam the exhaust makes a draft on the fire, and increases the heat through the flues, consequently the water pumped into the boiler is heated rapidly, and does not cause much contraction. When the throttle is shut off, very little heat passing through the flues, the water is not heated, and the boiler is cooled. Should it be absolutely necessary to pump with the throttle shut off, its effects will be partially overcome by putting on the blower full blast.

COUNTERBALANCING.

Counterbalancing is a matter upon which very little advice can be given, although proper attention to it is of the first importance. Should the engine not ride steady, and there is reason to believe the trouble is with the counterbalance, the Engineer must experiment with it, using his judgment in changing it one way or the other, until it is found to be right. He should always be certain, however, that the wedge between the engine and tender is set up snug. It is with this as with many other matters about an engine, in which the judgment of the thoughtful engineer will generally suggest the best mode of procedure. He should satisfy himself that he is right and then go ahead.

EXPANSION AND EXPANSION BRACES.

When an engine is fired up and under a good pressure of steam, the boiler and engine frame are longer than when they are cold. Hence, in order that the boiler shall maintain its proper position in the frame under all conditions, expansion braces are provided, and should be looked after with care, so that both sides will expand and contract alike. It is easy to ascertain the exact amount of expansion by marking the frame close to the expansion brace when the engine is hot, and noting the marks when it gets cold. Both sides of the engine should be accurately measured in this respect, and if the expansion is unequal, a remedy must be provided, as sooner or later a leaky throat sheet will result, the engine will be out of tram and ride hard, and it will be impossible to keep the side rods adjusted to the proper length, while in frosty weather there will be increased liability to break the rods. Engines frequently ride so hard as to endanger the health of the Engineer who attempts to run them, when perhaps the fault is solely occasioned by their being out of tram, or by unequal expansion. In such cases proper tramming and relief to the expansion braces usually cure the trouble, and even when locomotives are placed in the shop for general repairs, it is well that Engineers should see that these matters are not overlooked.

ACCIDENTS AND TEMPORARY REPAIRS.

Brief directions in reference to the best mode of management in sudden emergencies, when a breakdown occurs from any cause, will prove valuable, and enable the engineer, by making minor repairs on the spot, to reach home with comparatively little difficulty. Of course no directions can be given that will cover every case of sudden breakage, but those injuries most frequently occurring to the engine in case of accidents may be foreseen and at least partially remedied.

Should the locomotive run off the track, the first duty of the Engineer, if he is personally uninjured, will be to note the position of his engine, and see whether or not the water is withdrawn from any part of the fire box. If it is, the fire must be extinguished immediately, or the fire box will be ruined. Either earth, snow or coal, as is most convenient, may be used to quickly smother the fire.

If a driving spring be broken on the road, as is very often the case, place a hard wood block under the end of the equalizer opposite the unbroken spring. The block should be large enough to keep the equalizer level. To get it in place, run your engine on a block of wood that will raise it off the box containing the broken spring, and then with the assistance of a bar insert the block, letting it rest on the frame and under the equalizer. The blocking will bring the weight on the remaining spring. After removing the broken spring, the Engineer will be ready to proceed. The same blocking will be necessary if the spring hanger should break. If the equalizer breaks, the blocking must be under the frame on top of both driving boxes, taking out both springs and hangers, and the pieces of the equalizer, if possible, so that they will not catch in the driving wheels when the engine moves. If the engine has the old fashioned stirrup or strut, there will be little difficulty in blocking. It will be prudent to carry in the tender two ordinary rubber car springs of the right size for use in case of emergency. With these it will only be necessary to run the engine on a block, one wheel at a time, taking care to have the block large enough

ACCIDENTS AND TEMPORARY REPAIRS. 77

to raise the engine sufficiently to allow room to insert the rubber block. Place pieces of board below and above the rubber to give it good bearing. When one box is right, move the engine off the blocking and fix the other the same way. In an engine with the new style of driving boxes, having ears cast upon them, pieces of wood must be used for blocking instead of rubber. The engine should be blocked up as nearly level as possible; with rubber blocks there will be no danger of breaking the frames, but with wooden ones the Engineer must run slowly and carefully, though there will probably be no necessity of leaving the train behind. It is always well to be provided with extra spring hangers, as they will sometimes save a good deal of trouble.

If the tire of the driving wheel is broken or comes off, or loosened so as not to be safe, disconnect that side of the engine. If it be the forward driver, take down the main rod and the side rods. If one side rod be removed the other must also be taken down, as the engine should never be run with one side rod, for it will almost inevitably be broken, especially if the engine has four drivers. If the back driving wheel tire is broken, take down both side rods, block or sling up the wheel with the broken tire by running it on a block high enough to run it under the boxes on the pedestal binder so that the wheel will clear the track. Then run slowly and carefully without the train, reducing the speed still more when passing over frogs or crossing plates. The blocking should be no longer than the box, so that it will clear every thing.

When the driving-shaft breaks outside the box, block up under the box sufficiently to bring the box at an equal height with the other side.

Should a break occur which will render it necessary to disconnect the main rod, the piston must be secured by moving the cross-head to the extreme back end of the guides. Disconnect the valve stem from the rocker-arm and pull the valve clear back so as to cover the back steam-port, and screw up hard the glands to the stuffing-box and piston to prevent them from moving. This will answer until the Engineer is out of the way of approaching trains, and afterward a piece of board may be tied in between the guides. The advantages of this method of disconnecting the rod are obvious. Placing the cross-head at the back end of the guides brings the piston-head to the back end of the cylinder, and pulling the valve back opens the forward steam-port, so that when the throttle is opened, the steam fills the forward end of the cylinder, and the piston is held solid and cannot move while the throttle is open.

In case of a broken cylinder, cover the ports with the valve by moving the reverse lever until the rocker-arm stands as nearly plumb as possible. If this will not bring the rocker-arm plumb, disconnect the valve stem from the rocker-arm and place the valve where the rocker-arm would bring it if it were plumb. Then disconnect the valve-stem in the middle and screw up the stuffing-box gland hard, so that it will hold the valve in the required position.

The method of treating broken truck shafts will vary of course according to the character of the truck. The engineer must use his judgment. The truck must be well chained up and secured so that it will not swing. Should it be a tender truck, however, place a tie across the top of the tender and chain to both ends. No one rule can be laid down with regard to engine trucks that would apply to all the different styles in use.

A broken tumbling shaft is temporarily remedied by placing a piece of wood inside the links on the top of the link block, of sufficient length to permit the use of steam enough to start the train either on a level or a grade. This will save the engine much needless labor. Should one link-lifter break, use the same remedy, and if the engine can draw the train working at twelve or fourteen inches, place the reverse lever in the notch desired, lift the link with the broken lifter to the height of the other link and place the piece of wood in the link as already described. The link blocks should be kept as nearly as possible at the same height, so that the engine will work square, and backing up should be avoided as much as possible, for when that is done, the piece of wood in the link must be lengthened.

MISCELLANEOUS SUGGESTIONS.

Slippery Rails.

An engine should never be worked so as to use the steam too expansively when the condition of the track is such that the driving wheel will slip badly. Owing to the increase of lead in the valve under such conditions, there will be a greater variation of steam in the cylinder than there would be at full stroke. When the engine is working with six or eight inches of steam, the valve, by reason of the lead, gives an opening to the steam port before the piston reaches the dead center. By the time the piston has passed the center point, and before it goes back far enough to give the steam leverage on the crank pin, there will be an increased quantity of steam in the cylinder nearer boiler pressure. By the time the engine secures a leverage on the crank pin, the pressure is so great and so quickly applied that there is no relief for the drivers but in slipping. Working with a full stroke and a light throttle, however, enables the engine to receive the steam easily and with a uniform pressure, and there will be less likelihood of slipping. This explanation further illustrates the statement heretofore made that lead on the valve is a benefit to the engine. With full stroke and light throttle very good time may be made with patience, much better than if the engine is worked expansively.

Sanding the Rails.

It is a common fault among Engineers to use too much sand to remedy the slipping of the drivers. It should be remembered that all that is used in excess of the quantity required to make the engine adhere to the rail is a disadvantage. After the drivers pass over it the car wheels pick it up one after the other, until every wheel on the train is more or less encumbered with it, and the draft of the train is sometimes increased to such an extent that the engine cannot overcome it. Engineers are apt to get out of patience when the drivers persist in slipping while ascending a grade, and thoughtlessly open the sand-box valves and cover both rails with sand. The situation is not improved when it is found that there is not power enough in the engine to move the clogged train. Such a predicament is avoided by using the sand as sparingly as if it were a very costly article. Of course a sand-box filled with sand is indispensable on a locomotive, and often convenient in case of emergency. It should therefore be kept in perfect order, and used judiciously. It is often needed as much to hold or stop a train as to prevent slipping.

Contraction of the Flues.

It has already been stated that pumping cold water into the boiler after shutting off the throttle will cause sudden contraction in the flues. The same result will ensue if the furnace door is allowed to remain wide open, and both flues and fire-box will contract. If it is desirable to reduce the pressure of steam it is better to use a little more pump, and direct the fireman to cover the fire a little, leaving the door on the latch or opened a trifle. Thus steam may be gradually reduced without injury to the boiler or its parts. Where heavy trains are to be hauled, or steep grades surmounted, with the additional drawback of bad water, the effects of sudden contraction are very soon observable in the condition of the boiler.

Smoke Stack and Draft.

Smoke stacks differ according to the ideas of different builders, and also according to the quality of coal to be used. There is very little difference, however, in the general principle of their construction. The draft is usually controlled by the petticoat or draft pipe, whose exact position can only be determined by experiment. The draft pipe should be two inches less in diameter than the hole in the saddle to the smoke stack, with a twenty-two or twenty-four inch flare at the bottom. It should be set level with the short tips on the exhaust pipes, with the top about two inches from the top of the arch, after which experimenting will show when it is in the right position to make steam properly. In some engines it will be better to let the draft pipe run up into the stack eight or ten inches.

Lining up Cross-heads.

Where the old fashioned gibb or gibb cross-head is used, the lining up is done by taking off the outside plate and also the inside plate, carrying the lug to which the pump plunger is fastened, and then, with the aid of a pair of dividers, working from the center marks on the cross-head. It will easily be perceived where the liners are wanted in order to keep the center of the cross-head in the center of the guides. In case the cross-head is a solid one, the lining must always be done at the end of the guides, and on the block between them. If the lining is done at the center of the guides, and they should happen to be sprung, they will bind at the end, and perhaps break something.

Position of the Engine with reference to the Driving Boxes.

Should the engine get too low on the driving boxes owing to the springs losing their set, allowing the frame to strike on the boxes, it will be necessary to raise the engine by shortening the hanger, using washers made for the purpose. On the other hand, should the engine become too high the driving box will strike the pedestal binders. In that case the hangers must be lengthened, which will let the engine down to its proper place, the Engineer being governed as to distance by the variation in the hangers.

Elevation of Boiler.

The boiler of the locomotive should be elevated a little higher at the front end than at the back, enough so that it will be perceptible to the eye. This situation will bring the greatest body of water around the fire-box, and where it will receive the greatest possible amount of heat. The front end of the engine may be raised by inserting a ring made in two halves in the female casting around the center pin, care being taken not to raise the male casting out of the female casting by putting in too many rings.

The Head Light.

The head light should always be kept neat and clean, and no night trip should be made without it if possible. If it will not burn, rather than use none at all, tie a hand lamp on the brackets. The head light is not designed so much to assist the vision of the Engineer, as to enable others to see the approaching engine. A good light is really a pleasant companion for the Engineer during the hours of the night, and very little time or trouble is required to keep it in good condition.

Heated Pumps.

When the check-valve leaks by reason of its being too small, and becomes cocked in the cage, or something gets under it, the pump will get hot and will not work. To remedy it, shut off the flow of water from the pump, open the pet cock and put the heater on full head. Let the pet cock remain open a few seconds and then shut it for a little while. Then shut the heater off, letting on the flow of water, and try the pet cock. If this process does not set the pumps working the first time, try several times. Jarring the check-casting lightly with a hammer, will in nearly every case remove the trouble and render the check tight. Two reasons may be given for using the heater to get the check down. 1st. By shutting off the flow of water from the pump and putting on the heater, the pressure below the water line is less than above, and the check being below the water line when the heater is put on, the cold water in the feed pipe is forced through the pump into the boiler, and coming in contact with the heated part of the pump, will reduce the expansion. 2d. It will raise the check and force any ordinary obstacle out of the cage into the boiler, and also, the expansion being reduced, remove the obstacles that may have caused the leakage in the valve. The valve will bottom or seat, and the pump will immediately commence working.

The Pumps in Freezing Weather.

During cold weather, when the Engineer is obliged to shut off the pump while in motion, the pet cock may be left wide open, and neither the pump, the valves or the pet cock will freeze. He will thus avoid the trouble of using the heater, as well as the liability of getting the check up, while the pump is always in readiness for work when wanted. This applies to the pump in constant use, while the heater may be used to the other pump to prevent its freezing.

A Laboring Engine.

It will occasionally happen that there will be a pound about the engine when all efforts to ascertain the cause will be baffled. When that is the case, see if the piston is not loose in the follower, making what is sometimes termed a loose piston. Sometimes, also, the bolts in the splice to the frame will be cut, and still remain in their places, leaving the Engineer to believe that every thing is all right. These matters should be looked after.

GENERAL REMARKS.

At this point it is deemed advisable to say something regarding the duties and responsibilities of the Engineer. Much has already been well said on this subject in the columns of the Engineer's Journal, but the topic is not by any means exhausted. No work designed to acquaint the Engineer with the duties of his profession would be complete did it not also aim to inspire him with a spirit of emulation as well as determination to so bear himself as a man, that his membership in the Engineering profession will reflect upon it added honor and credit. Every Engineer should be sensible of the great responsibility he assumes when he steps upon the foot-board, a responsibility, it may be said, which belongs to very few of the ordinary occupations of life. The keenest watchfulness and the most unremitting attention is required of him, at all times, in the performance of his duties, that he may be at any moment ready to meet trying emergencies, when perhaps all the skill and judgment he possesses may be called into requisition. He may often be called upon to decide, on the instant, matters upon which the gravest results depend, and which, perhaps, involve the safety of hundreds of human lives. In such cases a quick, unerring judgment and a cool head are of inestimable value, and the Engineer who possesses them at command, has that upon which an enviable reputation may be based. The Engineer usually has but slight warning when an accident takes place. But the instant which may be assumed to exist between the discovery that danger is imminent and the presence of the actual peril, may afford opportunity either to ward off the blow, or mitigate its effect. How well that opportunity is improved rests entirely with the man on the foot-board. And it is very certain that it will not be improved if that man is habitually careless or inattentive. It is true that men are sometimes overworked, especially on our larger railroads. The consequent fatigue may cause a relaxation of vigilance on the road, and account for many casualties attributed to sheer carelessness.

But even physical weariness is no excuse for carelessness, and the sense of responsibility which should be ever-present ought to suffice to keep the eye steadily forward and vigilance unrelaxed. Railroad officials ought, however, to be held strictly responsible, when they press more work upon their Engineers than they are able to perform, and when it is done and serious accidents result, as they will inevitably, the public should be made aware of the truth and the blame placed where it belongs. Meanwhile Engineers should steadily combat excessive overwork, and neither permit their employers to persuade or force them into it or be too willing to accept it because of the added amount of money it will bring on pay day. Every Engineer ought to make it an inflexible rule never to run a locomotive over the road when he feels reasonably doubtful of his physical capacity to complete the journey satisfactorily. In honesty to himself and to his employers he should refuse under such circumstances, and should remember that no master mechanic or superintendent would be willing to take from his shoulders the responsibility for any accident that might occur. His manliness would be respected also by those still higher in authority, and he would have the additional satisfaction of feeling that he had performed a manifest duty to his family and to the community.

Furthermore, Engineers sometimes fail because they are not always particular to discharge their whole duty strictly in accordance with orders. This is a serious fault, and one which, if persisted in, is sure to bring trouble. The smallest point should never be left undone or unattended to. In plain terms, Engineers should always mind their own business, keep their own counsel, and be cautious with regard to all their words and acts, remembering that others have fine feelings as well as they, and do not like to have them trampled upon. He should respect every man's feelings and neither commit himself nor place himself in the power of any person. By such a course he will preserve his own dignity and independence, and be able to meet every man on his own level. These qualities, joined with courtesy to all with whom he is brought in contact, will win friends and establish his character and reputation. On the other hand, neglect and careless performance of duty, surly and overbearing manners, and heedless and injudicious conversation or conduct with bosses and others, will surely result in trouble and misfortune. The Engineer guilty of these faults will ultimately cease in a measure to be conscious of them, and though he gets into trouble, will fail to ascribe it to its true cause, and will perhaps imagine, when he is suffering the consequences, that he is suffering unjustly. Such men are the first to feel the effect of dull times, and often a mere pretense is sufficient to render their positions vacant. It is a sad fact that there are many Engineers of this class, and perhaps they will continue to be more or less numerous. At all events, let the young Engineer who reads these words resolve that he will not be of the number.

There may be those inclined to make light of these remarks, but it is obvious that good morals, integrity of character and clear understanding are peculiarly valuable to an Engineer, who cannot be too circumspect in his conduct or too strict in the performance of the duties imposed upon him. Every Engineer, therefore, who strives to excel in this regard, does his part toward gaining the good opinion and respect of the community for his dangerous and responsible profession, and materially enhances his own welfare. Nothing worth keeping will in any sense be lost to the Engineer by an endeavor to live a life of truth, justice, sobriety and morality, and consistently adhering to the golden rule, resolutely excluding malice and jealousy in his daily intercourse with brother Engineers.

SPECIAL ADVICE TO ENGINEERS AND FIREMEN.

Having explained the principal points of the locomotive, and pointed out the duties of the locomotive fireman and engineer, as such, it is perhaps not out of place for one whose observation and experience extend over a period of more than a score of years in railroad life, to call attention to another matter which may not come strictly within the scope of this treatise, but which is nevertheless of the highest importance.

While it is true that upon our railroads are men of the noblest and most generous impulses, still you have undoubtedly observed the alarming extent to which evil habits prevail among them. We will presume that you are an upright, moral man. If so, let your influence be felt wherever your presence may be. A look from you may so rebuke your profane companion that he will cease to swear where you can hear, and your persuasion may be effectual in inducing another to abandon the intoxicating cup. You will be met by many arguments in defense of wrong. One thinks it is manly to swear or to drink at the bar with a friend — thinks he cannot be a man among men if he does not. Tell him his position is a wrong one, tell him that down in the deep of his own heart is the consciousness that he himself will most readily trust the man who pronounces the name of God reverently, and whose breath is not tainted with intoxicating drinks. Tell him that such habits degrade and unfit him for companionship with the pure and the good. Tell him that the traveling public cherish the highest respect for all who act the gentleman and that no man is regarded as gentlemanly who uses profane language, or is known to be under the influence of strong drink. Tell him that any one of the ordinary habits of vice indulged makes it very easy to fall into others, diminishes self-respect, self-control and makes him less a man, and this he has no right to become. His friends will hold him responsible for what he might have been. Tell him that his relation to the company binds him to scrupulously guard and protect their interest, and this he cannot do without a clear head, and honest heart. Tell him that men of uprightness of character are the ones above all others to be intrusted with important interests, and such are usually sought. If he has a family, show him how they will be influenced by his actions. His wife may be a jewel, and his children as sweet and lovable as any under the sun, but who will seek them out and introduce them into society, especially if they should first meet him, who so unjustly and falsely represents them? As a rule they rise and fall with him, and as no man can afford to sacrifice his family, tell him to be true to the noblest instincts of his nature, and the highest capacities of his being. Finally, tell him that while the externals of morality alone will not insure entrance into life, the practice of vice will effectually debar from it.

INDEX.

	PAGE.
Advice, special, to Engineers and Firemen	96–98
Accidents, duty of Engineer when they occur	75
mode of management in case of	75
Boiler, advantage of high water in	70
check hole in	41
effect of pumping cold water into, with throttle shut off	70
elevation of	86
should be water in, before starting fire	11
Blowing in cylinders and valves, how to detect and how to remedy	44–48
Brasses, obtaining centers of	83
rule for reducing	87
Cab, cleaning interior of	12
Center-marking, how done	31
Check-valve, mode of remedying leaky	87
Counterbalance, experimenting with	72
Cross-heads, mode of lining up	85
used to obtain starting point at extreme dead center	54
why placed on extreme forward center in setting eccentrics	63, 64
Cylinder, blowing in	44–48
clearance of	35
how steam is admitted in	49
mode of procedure when broken	79
turning packing rings in	44

	PAGE
Cylinder packing, directions regarding	44–48
grinding in	44, 45
loss of power when not tight	45, 46
Draft-pipe, dimensions, etc., of	84
Driving-shaft, mode of blocking, when broken	78
Driving-spring, broken, how to repair	75, 76
Driving-wheels, how they should be trammed	30, 31
Eccentrics, how placed in direct and indirect motion engines	60, 61
manner of setting, on the road	63–65
rule regarding throw of	60
Engine, both sides of, should be measured to ascertain the expansion	73
Engineer should ascertain when repairs to, are needed	22
Firemen's duties when starting on a trip with	11–13
housing, after a trip	14
how injured by pumping with throttle shut off	71
laboring, where to look for cause of	89
moving, around yards and stations	27
oiling	12
position of, to be first noted in case of leaving track	75
position of, with reference to the driving-boxes	85
remedy for riding hard	73, 74
worn out, assignment of young engineers to	28
Engineer, advice to, when on single track	20
duties of, on completion of trip	21, 22
duty of, when engine is off the track	75
general duties outlined	17–22
general remarks on the duties and responsibilities of	90–95

INDEX.

	PAGE
Engineer, indispensable requisites of	17
precautions to be observed by, in running over road,	18
preparations for a trip	17–19
should avoid blocking the road when in trouble	21
should closely watch for defects in engine	21, 22
should possess a good time-piece	17
should see that sufficient water is in boiler	19
should start his engine with care	18
when preliminary knowlege of, is best obtained	16
Engineers and Firemen, special advice to	96–98
Equalizer, broken, how repaired	76
Expansion braces, why provided	73
Expansion, exact amount of, should be ascertained	73
Extreme or dead center, how ascertained	56, 57
Fire, directions for starting	11
precautions to be observed in starting	11, 12
to be extinguished immediately when engine runs off the track	75
Fireman, advice to those seeking promotion	15, 16
duties on completion of trip	14, 15
duties when starting engine on a trip	11–13
duties while running over the road	13
general duties outlined	11
should manifest interest in his duties	13
Firemen, Engineers and, special advice to	96–98
Firing, when, should not be done	13, 14
Flues, contraction of	83
Freezing weather, mode of treatment of pumps in	88
letting fire go out in	21
Freight engines, beginners usually placed on, at first	27
Full port, when obtained with and without lead	54

PAGE.

General remarks on the duties and responsibilities of
    the Engineer 90–95
Go-ahead eccentric, throw of...................... 60, 61
Head-light, injunctions regarding................. 87
Heated pumps, remedy for........................87, 88
Hooks, eccentric, how equalized with throw of eccentric, 64
Hostlers.. 15
Jack, use of, in adjusting rods.................... 36
Journals, engine and tender, to be inspected...... 28
Lap, functions of................................. 50–52
    inside, advantages and disadvantages of........ 51, 52
    inside, why suitable for freight engines........... 51
    lead equivalent to............................... 55
Lead, how obtained 53
    on valve, explanation of........................ 53–55
    utility of....................................... 54, 55
Liners, use of, in adjusting rods and brasses........ 33–37
Link-lifter, mode of treating when broken............ 80
Locomotive, definition of 9
Long runs for water should be avoided.............. 20
Low water, results of............................. 11–19
Main rod, how to disconnect...................... 78, 79
Motion, direct and indirect....................... 61, 62
Oiling engine..................................... 12
Packing pumps.................................... 40
Passenger engines should have very little inside lap... 52
Pet cock, management of......................... 42, 43
    result of neglect of............................ 43
Piston, clearance and travel of.................... 35, 36
Pumping, importance of proper attention to, when on
    the road 69, 71

INDEX. 103

	PAGE.
Pump, packing of	40
should not be worked after throttle is shut off	69
when advisable to shut off	70, 71
Pumps, liners and rings should be watched	41, 42
general directions regarding	38–43
mode of treatment in freezing weather	88
overhauling	39
result of too hard packing	40
should be repacked when they commence to leak	40
should work freely and uniformly	38
Pump-valves, description of	39
how should be set	39
lift of	39
Rods, adjustment of	32–37
Rubber springs, should be kept on tender for emergencies	76
Sand, should be used sparingly on slippery rails	83
Sanding the rails, cautions regarding	82
Side and main rods, adjustment of	32–37
Side rods, adjusting to proper length	33, 34
keying up on pins	35
tramming and lining	33–35
Slippery rails, suggestions regarding	81
Smoke-stack, suggestions regarding	84
Special advice to Engineers and Firemen	96–98
Steam ports, description of	49–52
Tank, to be kept clean	40
Temporary repairs, directions for making	75
Throw of eccentric, rule regarding	60
Tire, mode of procedure when disabled	77, 78
Train, breaks in, to be avoided	27

	PAGE.
Tramming driving-wheels	30, 31
Tramming, explanation of	30
when and how it is done	30
Trucks, packing	28
Truck shafts, method of treating when broken	79, 80
Tumbling shaft, temporary remedy when broken	80
Valve, cavity in	50
explanation of lead on	53–55
how amount of lead of, is ascertained	59
laps of	50–52
travel of, how to detect obstruction to	68
utility of lead of	54, 55
Valve motion, course of procedure when out of order,	66–68
four points of	67
position of the engine when examining	67
trouble with, on the road, how remedied	66–68
Valves, how set	57, 58
philosophy and principles of	49, 50
Valve seat, should be above opening in cage	41
strainer to, should be air-tight	41
ports of	49
Water and fuel, directions to Fireman regarding	12
Water in boiler, management of	19
Wedges, setting up	32
Young Engineers, advice to	23–29
assignment of, to worn-out engines	28
suggestions to, when placed on a strange engine,	24–27
when worthy of promotion	28

www.ingramcontent.com/pod-product-compliance
Lightning Source LLC
Chambersburg PA
CBHW020154170426
43199CB00010B/1036